AI IN AGRICULTURE:
HOW TO USE AI IN FARMING

BY

HENRY E. PARKINS

COPYRIGHT PAGE

TABLE OF CONTENTS

INTRODUCTION

In recent years, the intersection of technology and agriculture has led to revolutionary changes in the way we cultivate crops, manage livestock, and sustainably produce food for our growing global population. At the forefront of this transformation lies Artificial Intelligence (AI), a powerful tool that offers unprecedented opportunities to optimize farming practices, enhance productivity, and mitigate the challenges facing the agricultural industry.

The book "AI in Agriculture: How to Use AI in Farming" delves into the intricate relationship between AI and agriculture, offering insights into how this transformative technology can be harnessed to address the complex and evolving needs of modern farming. From precision agriculture to supply chain optimization, AI holds the potential to revolutionize every aspect of agricultural production and management.

As we stand on the brink of a new era in agriculture, it is imperative to understand the fundamental principles of AI and its applications in farming. This book serves as a comprehensive guide, providing readers with a deep dive into the fundamentals of AI technology, its practical applications across different agricultural domains, and the challenges and opportunities associated with its implementation.

Through a combination of theoretical insights, practical case studies, and real-world examples, readers will gain a nuanced understanding of how AI can drive innovation, improve efficiency, and promote sustainability in agriculture. From crop management and soil health monitoring to livestock management and supply chain optimization, AI offers solutions to some of the most pressing challenges facing the agricultural industry today.

Moreover, this book explores the broader implications of AI adoption in agriculture, including ethical considerations, societal impacts, and the role of policymakers in fostering an enabling environment for technological innovation. By embracing AI-driven solutions, farmers, researchers, policymakers, and other stakeholders can work together to create a more resilient, efficient, and sustainable agricultural system for generations to come.

In the pages that follow, readers will embark on a journey through the cutting-edge landscape of AI in agriculture, exploring the potential, challenges, and transformative power of this groundbreaking technology. Whether you are a seasoned agricultural professional, a technology enthusiast, or simply curious about the future of farming, "AI in Agriculture: How to Use AI in Farming" offers valuable insights and inspiration for leveraging AI to unlock new possibilities in agriculture.

Definition of AI in Agriculture

Artificial Intelligence (AI) in agriculture refers to the application of advanced computational algorithms and techniques to optimize various aspects of farming practices and food production. At its core, AI empowers agricultural systems to analyze vast amounts of data, make intelligent decisions, and automate processes with unprecedented accuracy and efficiency.

In the context of agriculture, AI encompasses a diverse range of technologies, including machine learning, computer vision, robotics, and natural language processing, among others. These tools enable farmers, agronomists, and other stakeholders to gain deeper insights into crop health, soil quality, weather patterns, and market trends, ultimately enhancing decision-making and resource allocation throughout the agricultural value chain.

AI in agriculture encompasses a wide array of applications, from precision farming and crop monitoring to livestock management and supply chain optimization. By leveraging AI-driven solutions, farmers can optimize resource utilization, reduce waste, and mitigate risks associated with unpredictable environmental factors and market dynamics.

Moreover, AI holds the potential to revolutionize traditional farming practices by enabling autonomous machinery, smart sensors, and predictive analytics systems to work in tandem, creating a more connected and responsive agricultural ecosystem. Through continuous learning and adaptation, AI-driven agricultural systems can adapt to evolving conditions, optimize yields, and promote sustainable land management practices for the benefit of both farmers and the environment.

In essence, AI in agriculture represents a paradigm shift in how we approach food production and resource management, offering innovative solutions to the complex challenges facing the agricultural industry in the 21st century. As we navigate the complexities of a rapidly changing global landscape, AI has emerged as a powerful tool for driving efficiency, sustainability, and resilience in agriculture, paving the way for a more food-secure and environmentally sustainable future.

Importance and Relevance of AI in Modern Farming Practices

In today's rapidly evolving agricultural landscape, the importance and relevance of Artificial Intelligence (AI) cannot be overstated. As the global population continues to grow, along with increasing demands for food security, sustainability, and efficiency, farmers and agricultural stakeholders are faced with unprecedented challenges. In this context, AI emerges as a transformative force, offering innovative solutions to enhance productivity, optimize resource management, and address the complexities of modern farming practices.

Precision Agriculture: AI plays a pivotal role in enabling precision agriculture, a data-driven approach to farming that leverages advanced technologies to optimize crop yields while minimizing input costs and environmental impact. Through the integration of AI-driven tools such as sensors, drones, and satellite imagery, farmers can gather real-time data on soil health, weather patterns, and crop conditions, allowing for more targeted and informed decision-making at every stage of the growing season.

Data-driven Decision-making: In an era characterized by the proliferation of big data, AI empowers farmers to harness vast amounts of information and extract actionable insights to improve operational efficiency and profitability. By leveraging machine learning algorithms and predictive analytics, farmers can analyze historical data, identify patterns, and forecast future trends, enabling them to make informed decisions regarding planting schedules, irrigation strategies, pest management, and market opportunities.

Sustainable Agriculture: With growing concerns about environmental degradation, water scarcity, and climate

change, the imperative for sustainable agriculture has never been greater. AI offers a suite of tools and technologies to promote more sustainable farming practices, including optimized resource utilization, reduced chemical inputs, and improved soil health management. By enabling precision application of fertilizers, pesticides, and irrigation water, AI helps minimize waste and environmental impact while maximizing the long-term viability of agricultural ecosystems.

Enhanced Crop and Livestock Management:
AI-driven technologies such as computer vision and machine learning algorithms enable farmers to monitor and manage crops and livestock with unprecedented precision and efficiency. From detecting crop diseases and nutrient deficiencies to tracking animal health and behavior, AI-powered systems provide real-time insights into the health and well-being of agricultural assets, enabling proactive interventions

and improving overall productivity and profitability.

Supply Chain Optimization: In an interconnected global marketplace, efficient supply chain management is essential for ensuring timely delivery of agricultural products to markets and consumers. AI facilitates supply chain optimization through predictive analytics, demand forecasting, and logistics optimization, enabling farmers to streamline production processes, minimize waste, and respond more effectively to market fluctuations and consumer preferences.

Overview of the Book's Objectives and Structure

"AI in Agriculture: How to Use AI in Farming" is a comprehensive guide that explores the intersection of Artificial Intelligence (AI) and agriculture, offering insights into the transformative potential of AI-driven technologies in modern farming practices. The book aims to provide farmers, agronomists, researchers, policymakers, and other stakeholders with a holistic understanding of how AI can be leveraged to optimize agricultural production, enhance sustainability, and address the complex challenges facing the agricultural industry.

Objectives:

Educate and Inform: The primary objective of the book is to educate readers about the fundamentals of AI technology and its applications in agriculture. By demystifying complex concepts and providing real-world examples, the book aims to empower readers with the knowledge and insights needed to harness the power of AI in their farming practices.

Showcase Practical Applications: Through a series of case studies, examples, and practical insights, the book illustrates how AI can be applied across various domains of agriculture, including crop management, livestock monitoring, soil health analysis, and supply chain optimization. By showcasing successful implementations and highlighting best practices, the book seeks to inspire readers to explore new avenues for integrating AI into their own agricultural operations.

Address Challenges and Opportunities: The book acknowledges the challenges and barriers to AI adoption in agriculture, including issues related to cost, data privacy, infrastructure, and technical expertise. By addressing these challenges head-on and providing practical guidance, the book aims to help readers navigate the complexities of integrating AI into their farming practices and capitalize on the opportunities for innovation and growth.

Foster Collaboration and Knowledge Sharing: In addition to providing valuable insights and practical guidance, the book serves as a platform for fostering collaboration and knowledge sharing among stakeholders in the agricultural community. By bringing together diverse perspectives and experiences, the book

16

encourages dialogue, collaboration, and the exchange of ideas to drive collective action and positive change in the agricultural sector.

Structure:

Introduction: The book begins with an overview of the importance and relevance of AI in modern farming practices, setting the stage for the exploration of key concepts and themes.

Fundamentals of AI Technology: The second section delves into the fundamental principles of AI technology, providing readers with a foundational understanding of machine learning, computer vision, robotics, and other AI-driven technologies relevant to agriculture.

Applications of AI in Farming: The heart of the book explores the diverse applications of AI in agriculture, including crop management, livestock monitoring, soil health analysis, and supply chain optimization. Each chapter examines specific use cases, practical implementations, and real-world examples to illustrate the transformative potential of AI in modern farming practices.

Case Studies and Examples: The book features a series of case studies and examples that highlight successful AI implementations and best practices in agriculture. These real-world examples provide readers with insights into how AI can be leveraged to address specific challenges and opportunities across different agricultural domains.

Implementing AI in Agriculture: The penultimate section offers practical guidance and actionable insights for integrating AI into farming operations. From selecting the right technologies to overcoming implementation barriers, this section provides readers with the tools and knowledge needed to successfully deploy AI solutions in their own agricultural practices.

Future Trends and Developments: The final section explores emerging trends and developments in AI and agriculture, offering insights into the future of farming and the potential impact of AI-driven technologies on the agricultural industry.

CHAPTER 1

FUNDAMENTALS OF AI TECHNOLOGY

In the realm of agriculture, understanding the fundamentals of Artificial Intelligence (AI) technology is essential for unlocking its transformative potential and harnessing its capabilities to optimize farming practices. This section of the book aims to provide readers with a foundational understanding of key AI concepts and principles relevant to agriculture, including machine learning, computer vision, robotics, and natural language processing.

Machine Learning: At the core of AI technology lies machine learning, a branch of artificial intelligence that enables systems to learn from data and improve their performance over time without being explicitly programmed. In the context of agriculture, machine learning algorithms can analyze vast amounts of agricultural data, including soil composition, weather patterns, crop health, and pest infestations, to identify patterns, make predictions, and inform decision-making processes.

Computer Vision: Computer vision is another critical component of AI technology that enables machines to interpret and analyze visual information from images or videos. In agriculture, computer vision algorithms can be used to assess crop health, detect diseases and pests, monitor plant growth, and identify weeds. By leveraging computer vision technology, farmers can gain valuable insights into the condition of their crops and make informed decisions about resource allocation and management.

Robotics: Robotics plays a crucial role in the automation of various agricultural tasks, ranging from planting and harvesting to sorting and packaging. AI-powered robots equipped with sensors, cameras, and actuators can perform repetitive tasks with precision and efficiency, reducing labor costs and increasing productivity. In agriculture, robotic systems can be used for tasks such as precision planting, weed removal, and fruit picking, enabling farmers to optimize their operations and maximize yields.

Natural Language Processing (NLP): Natural language processing is a branch of AI technology that focuses on enabling machines to understand, interpret, and generate human language. In the context of agriculture, NLP can be used to analyze text-based data sources such as research papers, agricultural reports, and social media posts to extract valuable insights and trends. NLP algorithms can also be used to develop chatbots and virtual assistants that provide farmers with real-time information and support.

Data Analytics and Predictive Modeling: AI technologies enable farmers to harness the power of data analytics and predictive modeling to make data-driven decisions and optimize agricultural operations. By analyzing historical data and trends, predictive modeling algorithms can forecast future outcomes, such as crop yields, market prices, and weather patterns, helping farmers anticipate risks and opportunities and plan accordingly.

Explanation of Artificial Intelligence (AI)

Artificial Intelligence (AI) represents a transformative field of computer science that focuses on creating systems capable of performing tasks that typically require human intelligence. In essence, AI enables machines to simulate human cognitive functions such as learning, reasoning, problem-solving, perception, and decision-making.

At its core, AI encompasses a broad range of techniques, algorithms, and methodologies designed to enable machines to understand, analyze, and respond to complex data inputs in a manner that mimics human intelligence. These techniques include:

Machine Learning: Machine learning is a subset of AI that focuses on enabling computers to learn from data without being explicitly programmed. Through the use of algorithms and statistical models, machine learning systems can analyze large datasets, identify patterns, and make predictions or decisions based on the information they have processed. In agriculture, machine learning algorithms can be used to analyze data related to crop health, soil composition, weather patterns, and market trends, enabling farmers to make informed decisions about resource allocation, pest management, and crop selection.

Deep Learning: Deep learning is a subset of machine learning that focuses on enabling computers to learn from data representations, such as neural networks, that are inspired by the structure and function of the human brain. Deep learning algorithms are particularly well-suited for tasks involving large amounts of unstructured data, such as image recognition, natural language processing, and speech recognition. In agriculture, deep learning algorithms can be used to analyze satellite imagery, drone footage, and sensor data to monitor crop health, detect diseases, and identify pest infestations with a high degree of accuracy.

Computer Vision: Computer vision is a subfield of AI that focuses on enabling machines to interpret and analyze visual information from images or videos. Through the use of algorithms and techniques such as image classification, object detection, and image segmentation, computer vision systems can identify objects, patterns, and anomalies within visual data. In agriculture, computer vision algorithms can be used to monitor crop growth, detect weeds, and assess soil conditions, enabling farmers to make data-driven decisions about irrigation, fertilization, and pest control.

Natural Language Processing (NLP): Natural language processing is a subfield of AI that focuses on enabling machines to understand, interpret, and

generate human language. Through the use of algorithms and techniques such as text classification, sentiment analysis, and named entity recognition, NLP systems can analyze and extract valuable insights from textual data sources such as research papers, social media posts, and agricultural reports. In agriculture, NLP algorithms can be used to analyze weather forecasts, market reports, and research articles to provide farmers with real-time information and recommendations.

Types of AI Relevant to Agriculture

In the dynamic landscape of agriculture, various types of Artificial Intelligence (AI) technologies are reshaping traditional farming practices and offering innovative solutions to enhance productivity, sustainability, and efficiency. The integration of AI in agriculture encompasses a diverse array of technologies, each with its unique capabilities and applications. In "AI in Agriculture: How to Use AI in Farming," we explore the following types of AI relevant to agriculture:

Machine Learning: Machine learning lies at the heart of AI applications in agriculture. This branch of AI focuses on enabling systems to learn from data, identify patterns, and make decisions without explicit programming. In agriculture, machine learning algorithms analyze vast datasets related to soil composition, weather patterns, crop health, and market trends to optimize farming practices, predict crop yields, and enhance decision-making processes.

Computer Vision: Computer vision is a critical AI technology that enables machines to interpret and analyze visual information from images or videos. In agriculture, computer vision systems can analyze drone footage, satellite imagery, and field photographs to monitor crop growth, detect diseases, identify weeds, and assess soil conditions. By providing real-time insights into crop health and environmental conditions, computer vision

27

technology empowers farmers to make informed decisions about irrigation, fertilization, and pest control.

Robotics: Robotics plays a significant role in automating various agricultural tasks and enhancing operational efficiency on the farm. AI-powered robotic systems equipped with sensors, cameras, and actuators can perform tasks such as planting, harvesting, weeding, and sorting with precision and consistency. In addition to reducing labor costs and improving productivity, robotic systems enable farmers to optimize resource utilization and minimize environmental impact, contributing to sustainable agricultural practices.

Natural Language Processing (NLP): Natural language processing is a branch of AI that focuses on enabling machines to understand, interpret, and generate human language. In agriculture, NLP technologies can analyze textual data sources such as research papers,

agricultural reports, and social media posts to extract valuable insights, trends, and sentiments. NLP algorithms can also power virtual assistants and chatbots that provide farmers with real-time information, recommendations, and support, enhancing communication and decision-making processes on the farm.

Predictive Analytics: Predictive analytics is an AI-driven approach to forecasting future outcomes based on historical data and statistical models. In agriculture, predictive analytics algorithms can analyze historical weather data, crop yields, market prices, and other relevant factors to predict future trends, risks, and opportunities. By providing insights into factors such as crop growth, pest infestations, and market demand, predictive analytics enables farmers to anticipate challenges, optimize resource allocation, and maximize profitability.

Integration of AI with Agricultural Systems

In "AI in Agriculture: How to Use AI in Farming," we delve into the seamless integration of Artificial Intelligence (AI) with agricultural systems, showcasing how AI technologies intersect with various aspects of farming practices to optimize productivity, sustainability, and efficiency. The integration of AI with agricultural systems encompasses the following key aspects:

Data Collection and Analysis: AI technologies facilitate the collection, aggregation, and analysis of vast amounts of agricultural data from diverse sources, including sensors, drones, satellites, weather stations, and farm machinery. By harnessing AI-powered analytics platforms, farmers can gain valuable insights into soil health, crop growth, weather patterns, pest infestations, and market trends, enabling data-driven decision-making and proactive management strategies.

Precision Agriculture: AI enables the implementation of precision agriculture techniques, which involve the precise application of inputs such as water, fertilizers, pesticides, and herbicides based on real-time data and analysis. Through the integration of AI-driven technologies such as machine learning, computer vision, and robotics, farmers can optimize resource utilization, minimize waste, and maximize yields while minimizing environmental impact.

Automation and Robotics: AI-powered robotics and automation systems play a pivotal role in streamlining agricultural operations and reducing labor-intensive tasks on the farm. From planting and harvesting to sorting and packaging, AI-driven robotic systems equipped with sensors, cameras, and actuators can perform tasks with precision, consistency, and efficiency, freeing up human resources for more strategic and value-added activities.

Decision Support Systems: AI-driven decision support systems provide farmers with actionable insights, recommendations, and predictions to guide decision-making processes across various aspects of agricultural management. By analyzing historical data, weather forecasts, soil conditions, and market dynamics, decision support systems help farmers optimize planting schedules, irrigation strategies, pest control measures, and crop rotation plans, resulting in improved productivity and profitability.

Supply Chain Optimization: AI technologies enable the optimization of agricultural supply chains by facilitating real-time monitoring, tracking, and management of farm-to-fork processes. Through the integration of AI-driven logistics platforms, farmers, distributors, and retailers can optimize inventory management, transportation routes, and delivery schedules, ensuring timely and efficient delivery of agricultural products to markets and consumers.

Continuous Learning and Adaptation: AI-powered agricultural systems leverage machine learning algorithms to continuously learn from data, adapt to changing conditions, and optimize performance over time. By analyzing feedback loops and historical outcomes, AI-driven systems can refine predictive models, improve decision-making algorithms, and enhance overall system performance, enabling farmers to achieve greater efficiency, resilience, and sustainability in their operations.

CHAPTER 2

APPLICATIONS OF AI IN FARMING

In "AI in Agriculture: How to Use AI in Farming," we explore a myriad of applications where Artificial Intelligence (AI) is transforming traditional farming practices and revolutionizing the agricultural industry. The integration of AI technologies across various domains of agriculture offers innovative solutions to enhance productivity, sustainability, and efficiency. Some key applications of AI in farming include:

Precision Agriculture: AI enables the implementation of precision agriculture techniques, which involve the precise management of inputs such as water, fertilizers, pesticides, and herbicides based on real-time data and analysis. Through the integration of AI-driven technologies such as sensors, drones, and machine learning algorithms, farmers can monitor soil

34

moisture levels, assess crop health, detect pest infestations, and optimize resource allocation, resulting in improved yields and reduced environmental impact.

Crop Monitoring and Management: AI-powered computer vision systems analyze satellite imagery, drone footage, and field photographs to monitor crop growth, detect diseases, identify weeds, and assess soil conditions. By providing real-time insights into crop health and environmental conditions, AI technologies empower farmers to make informed decisions about irrigation, fertilization, and pest control, leading to enhanced crop quality and yield.

Livestock Monitoring and Management: AI-driven monitoring systems equipped with sensors and machine learning algorithms enable farmers to monitor the health, behavior, and productivity of livestock in real-time. By analyzing data related to animal movements, feeding patterns, and

physiological parameters, AI technologies can detect signs of illness, optimize feeding schedules, and improve overall herd management practices, ensuring the well-being and productivity of livestock.

Soil Health Analysis: AI-powered soil health analysis tools leverage machine learning algorithms to analyze soil samples and assess key parameters such as nutrient levels, pH levels, and microbial activity. By providing insights into soil fertility, structure, and composition, AI technologies help farmers make informed decisions about soil management practices, including nutrient application, soil amendments, and crop rotation strategies, leading to improved soil health and productivity.

Supply Chain Optimization: AI technologies facilitate the optimization of agricultural supply chains by enabling real-time monitoring, tracking, and management of farm-to-fork processes. Through the integration of AI-driven logistics platforms, farmers, distributors,

and retailers can optimize inventory management, transportation routes, and delivery schedules, ensuring timely and efficient delivery of agricultural products to markets and consumers.

Predictive Analytics: AI-driven predictive analytics models analyze historical data, weather forecasts, market trends, and other relevant factors to forecast future outcomes such as crop yields, market prices, and weather patterns. By providing insights into future trends and risks, predictive analytics enable farmers to anticipate challenges, optimize resource allocation, and make informed decisions about planting schedules, irrigation strategies, and market opportunities, leading to improved profitability and resilience.

Crop Management

In "AI in Agriculture: How to Use AI in Farming," crop management stands as a pivotal domain where Artificial Intelligence (AI) technologies revolutionize traditional farming practices. Crop management

37

encompasses a range of tasks aimed at optimizing crop health, yield, and quality throughout the growing season. Through the integration of AI-driven tools and techniques, farmers can enhance decision-making processes, mitigate risks, and maximize productivity across various stages of crop production. Some key aspects of AI-driven crop management include:

Precision Agriculture: AI enables the implementation of precision agriculture techniques, which involve the precise management of inputs such as water, fertilizers, pesticides, and herbicides based on real-time data and analysis. Through the integration of AI-driven technologies such as sensors, drones, and machine learning algorithms, farmers can monitor soil moisture levels, assess nutrient levels, detect pest infestations, and optimize resource allocation, resulting in improved yields and reduced environmental impact.

Predictive Analytics for Crop Yield and Disease Management:

38

AI-driven predictive analytics models analyze historical data, weather forecasts, satellite imagery, and other relevant factors to forecast future outcomes such as crop yields, disease outbreaks, and pest infestations. By providing insights into future trends and risks, predictive analytics enable farmers to anticipate challenges, optimize resource allocation, and make informed decisions about planting schedules, irrigation strategies, and pest control measures, leading to improved productivity and profitability.

Remote Sensing and Monitoring: AI-powered remote sensing technologies leverage satellite imagery, drone footage, and other remote sensing data sources to monitor crop growth, detect diseases, identify weeds, and assess soil conditions. By providing real-time insights into crop health and environmental conditions, remote sensing technologies enable farmers to make timely interventions, optimize irrigation schedules, and minimize yield losses, resulting in improved crop quality and yield.

Automated Irrigation and Nutrient Management: AI-driven irrigation and nutrient management systems analyze soil moisture levels, weather forecasts, and crop water requirements to optimize irrigation schedules and nutrient application rates. By automating irrigation and nutrient management processes, AI technologies help farmers conserve water, reduce nutrient runoff, and improve nutrient uptake efficiency, leading to improved crop health and yield while minimizing environmental impact.

Weed Detection and Management: AI-powered computer vision systems analyze images of fields to detect and classify weeds based on their visual characteristics. By identifying weeds early in the growing season, AI technologies enable farmers to implement targeted weed control measures, such as precision herbicide application or mechanical weed removal, reducing weed

competition and improving crop yield and quality.

Crop Health Monitoring and Disease Diagnosis:

AI-driven monitoring systems equipped with sensors and machine learning algorithms enable farmers to monitor crop health and diagnose diseases in real-time. By analyzing data related to leaf color, texture, and morphology, AI technologies can detect signs of stress, nutrient deficiencies, and disease symptoms, enabling farmers to take proactive measures to mitigate risks and preserve crop health and productivity.

Precision Agriculture and Sensor Technologies

In "AI in Agriculture: How to Use AI in Farming," precision agriculture emerges as a transformative approach to farming, enabled by the integration of advanced sensor technologies and Artificial Intelligence (AI) algorithms. Precision agriculture aims to optimize resource use,

41

enhance productivity, and minimize environmental impact by tailoring farming practices to specific field conditions and crop requirements. At the heart of precision agriculture lie sensor technologies and AI-driven analytics, which enable farmers to monitor, analyze, and respond to dynamic environmental factors in real-time. Some key aspects of precision agriculture and sensor technologies include:

Soil Sensors: Soil sensors play a crucial role in precision agriculture by providing real-time insights into soil moisture levels, nutrient levels, pH levels, and temperature. AI-driven soil sensor networks enable farmers to monitor soil conditions across their fields, identify areas of moisture stress or nutrient deficiency, and optimize irrigation and fertilization schedules accordingly. By ensuring that crops receive the right amount of water and nutrients at the right time, soil sensors help maximize crop yields while minimizing water and fertilizer usage.

42

Weather Sensors: Weather sensors measure atmospheric conditions such as temperature, humidity, wind speed, and precipitation levels. AI-driven weather monitoring systems analyze weather data in real-time, enabling farmers to anticipate changes in weather patterns, plan agricultural operations accordingly, and mitigate risks associated with adverse weather events. By incorporating weather forecasts into decision-making processes, farmers can optimize planting schedules, irrigation strategies, and pest management practices, leading to improved crop resilience and yield stability.

Drone Technology: Drones equipped with advanced sensors and imaging technologies offer farmers a bird's-eye view of their fields, enabling them to monitor crop health, detect pests and diseases, and assess field conditions with unprecedented accuracy and efficiency. AI-powered drone imagery analysis algorithms process drone footage to identify areas of crop stress, weed infestation, or disease outbreak, enabling farmers to take targeted corrective actions and minimize yield losses. By providing high-resolution, real-time data, drone technology enhances decision-making processes and enables proactive crop management strategies.

Satellite Imagery: Satellite imagery provides farmers with a comprehensive view of their fields at a macroscopic scale, enabling them to monitor crop growth, assess field conditions, and identify spatial variability across large agricultural landscapes. AI-driven satellite image analysis algorithms process satellite imagery to detect changes in vegetation health, soil moisture levels, and crop performance over time, enabling farmers to identify trends, patterns, and anomalies that may impact crop yields and profitability. By leveraging satellite imagery and AI-driven analytics, farmers can optimize resource allocation, prioritize management interventions, and improve overall farm efficiency.

Crop Health Sensors: Crop health sensors measure physiological parameters such as leaf chlorophyll content, photosynthetic activity, and canopy temperature, providing insights into plant health and stress levels. AI-driven crop health monitoring systems analyze sensor data to detect signs of nutrient deficiencies, water stress, pest infestations, and disease outbreaks, enabling farmers to implement targeted management strategies and optimize crop performance. By monitoring crop health in real-time, farmers can identify potential problems early, minimize yield losses, and maximize profitability while promoting sustainable agricultural practices.

Predictive Analytics for Crop Yield and Disease Management

In "AI in Agriculture: How to Use AI in Farming," predictive analytics emerges as a powerful tool for improving crop yield and disease management. By leveraging historical data, weather forecasts, satellite imagery, and other relevant factors, predictive analytics models enable farmers to anticipate crop yields, identify disease outbreaks, and make informed decisions to optimize agricultural practices. Here's how predictive analytics is applied in crop yield and disease management:

Yield Prediction Models: Predictive analytics models analyze historical crop yield data, weather patterns, soil conditions, and agronomic practices to forecast future crop yields. By identifying trends, patterns, and correlations in the data, these models enable farmers to anticipate potential yield fluctuations, plan production schedules, and optimize resource allocation to maximize yields. Yield prediction models also help farmers evaluate the impact of different management practices and environmental factors on crop productivity, enabling them to make informed decisions to improve overall farm performance.

Disease Outbreak Forecasting:

Predictive analytics models analyze historical disease incidence data, weather patterns, crop phenology, and environmental conditions to forecast the likelihood of disease outbreaks in agricultural crops. By identifying factors associated with disease development and spread, these models enable farmers to implement proactive disease management strategies, such as timely application of fungicides, crop rotation, and sanitation practices, to mitigate risks and minimize yield losses. Disease outbreak forecasting models also help farmers prioritize monitoring efforts and allocate resources more effectively to prevent or control disease outbreaks in their fields.

Pest Infestation Prediction:

Predictive analytics models analyze historical pest population data, weather patterns, crop phenology, and landscape features to predict the likelihood of pest infestations in agricultural crops. By identifying factors conducive to pest development and spread, these models enable farmers to implement early detection and intervention strategies, such as insecticide applications, biological control methods, and habitat manipulation, to prevent or mitigate pest damage and minimize yield losses. Pest infestation prediction models also help farmers optimize pest management practices and reduce reliance on chemical pesticides, promoting sustainable and environmentally friendly farming practices.

Climate Resilience Planning:

Predictive analytics models analyze historical climate data, climate change projections, and crop performance under different climatic scenarios to assess the resilience of agricultural systems to climate variability and change. By simulating future climate scenarios and their potential impacts on crop productivity, these models enable farmers to identify adaptation strategies, such as crop diversification, adoption of drought-tolerant varieties, and modification of planting dates, to mitigate risks and maintain farm profitability in the face of changing climatic conditions. Climate resilience planning models also help farmers prioritize investments in climate-smart technologies and infrastructure to enhance the resilience and sustainability of agricultural production systems.

Livestock Monitoring and Management

In "AI in Agriculture: How to Use AI in Farming," livestock monitoring and management emerge as critical aspects of modern agricultural practices. With the integration of Artificial Intelligence (AI) technologies, farmers can monitor the health, behavior, and productivity of livestock in real-time, enabling proactive management strategies to enhance animal welfare and optimize production efficiency. Here's how livestock monitoring and management are facilitated through AI in agriculture:

Health Monitoring: AI-driven monitoring systems equipped with sensors

and wearable devices enable farmers to monitor the health status of individual animals and herd populations in real-time. These sensors can measure physiological parameters such as body temperature, heart rate, respiratory rate, and activity levels, providing insights into the health and well-being of livestock. AI algorithms analyze sensor data to detect early signs of illness, injury, or stress, enabling farmers to intervene promptly with appropriate veterinary care and management practices to prevent disease outbreaks and minimize production losses.

Behavior Analysis: AI-powered behavioral monitoring systems analyze patterns of movement, feeding behavior, and social interactions among livestock to assess animal welfare and identify signs of distress or discomfort. By monitoring changes in behavior over time, AI algorithms can detect deviations from normal behavior patterns, such as reduced activity levels or changes in social dynamics, which may indicate underlying health problems or environmental

50

stressors. Behavioral analysis enables farmers to implement targeted interventions, such as adjustments to feeding regimens, environmental conditions, or handling practices, to promote optimal welfare and performance in livestock populations.

Reproductive Management: AI technologies support reproductive management programs by enabling farmers to monitor estrus cycles, detect signs of heat, and optimize breeding strategies in livestock populations. AI-driven reproductive monitoring systems analyze physiological indicators such as hormone levels, body condition, and reproductive history to predict optimal breeding times and improve breeding efficiency. By identifying animals in estrus and coordinating mating activities accordingly, farmers can enhance reproductive success rates, minimize breeding-related costs, and optimize genetic selection programs to improve the overall quality and productivity of their livestock herds.

Feed Management and Nutritional Optimization: AI-driven feed management systems analyze dietary compositions, nutritional requirements, and feed intake patterns to optimize feeding strategies and nutritional regimens for livestock populations. By monitoring feed consumption, nutrient utilization, and growth rates in real-time, AI algorithms can identify dietary deficiencies, imbalances, or inefficiencies and recommend adjustments to feeding programs to optimize animal growth, performance, and health. Feed management systems also help farmers reduce feed wastage, minimize production costs, and improve feed conversion efficiency, leading to more sustainable and economically viable livestock production systems.

Disease Prevention and Control: AI technologies support disease prevention and control efforts in livestock populations by enabling early detection, rapid response, and targeted interventions to mitigate disease risks and minimize transmission pathways. AI-driven disease surveillance systems analyze health data, environmental conditions, and disease prevalence patterns to identify emerging threats and high-risk areas for disease outbreaks. By monitoring disease trends and risk factors in real-time, farmers can implement biosecurity measures, vaccination protocols, and quarantine procedures to prevent disease introduction and spread within livestock populations, safeguarding animal health and welfare while preserving farm productivity and profitability.

Automated Health Monitoring Systems

In "AI in Agriculture: How to Use AI in Farming," automated health monitoring systems stand out as crucial components in modern agricultural practices, particularly in livestock management. These systems integrate Artificial Intelligence (AI) technologies with sensor devices and data analytics to monitor the health status of livestock in real-time. By continuously collecting and analyzing data on physiological parameters and behavioral patterns, automated health monitoring systems enable farmers to detect early signs of illness, injury, or distress, facilitating timely interventions to safeguard animal welfare and optimize production efficiency. Here's how automated health monitoring systems are transforming livestock management:

Sensor Technologies: Automated health monitoring systems utilize a variety of sensor technologies to capture data on key physiological and behavioral indicators in livestock. These sensors may include wearable devices, implantable microchips, and environmental sensors deployed in barns and pastures. Sensors measure parameters such as body temperature, heart rate, respiratory rate, activity levels, rumination behavior, and feeding patterns, providing valuable insights into the health and well-being of individual animals and herd populations.

Real-Time Data Collection:

Automated health monitoring systems collect data from sensors in real-time, enabling continuous monitoring of livestock health and performance. Data streams from multiple sensors are aggregated and processed using AI algorithms to identify patterns, trends, and anomalies indicative of changes in animal health status. Real-time data collection allows farmers to detect health issues promptly and intervene before they escalate, minimizing the risk of disease outbreaks, production losses, and adverse welfare outcomes.

AI Analytics and Predictive Modeling:

AI-driven analytics and predictive modeling techniques are applied to sensor data to extract actionable insights and forecast health outcomes in livestock populations. AI algorithms analyze historical data, sensor readings, and environmental factors to identify early warning signs of health problems, such as deviations from normal baseline values or trends associated with specific health conditions. Predictive models can forecast the likelihood of disease outbreaks, assess the risk of individual animals, and recommend targeted interventions to prevent or mitigate health risks and optimize animal welfare.

Decision Support and Alerts:

Automated health monitoring systems provide farmers with decision support tools and real-time alerts to facilitate timely responses to health-related events in livestock populations. AI algorithms analyze sensor data and trigger alerts when predefined thresholds or patterns indicative of health problems are detected. Farmers receive notifications via mobile applications, dashboard interfaces, or automated messaging systems, allowing them to investigate potential issues, administer treatments, or implement management interventions promptly to address emerging health concerns and prevent further deterioration in animal health.

Integration with Management Practices: Automated health monitoring systems integrate seamlessly with existing livestock management practices and production workflows. Data generated by health monitoring systems can be integrated with electronic herd management systems, veterinary records, and production databases to support comprehensive decision-making and record-keeping processes. Integration with management practices allows farmers to track health trends over time, evaluate the effectiveness of interventions, and make data-driven adjustments to optimize herd health and performance.

Behavior Analysis for Optimal Livestock Care

In "AI in Agriculture: How to Use AI in Farming," behavior analysis emerges as a fundamental aspect of modern livestock management, facilitated by the integration of Artificial Intelligence (AI) technologies. Behavior analysis involves monitoring and interpreting the behavioral patterns of livestock to assess their health, well-being, and productivity. By leveraging AI-driven analytics and sensor technologies, farmers can gain valuable insights into animal behavior, enabling proactive interventions to optimize livestock care and management practices. Here's how behavior analysis contributes to optimal livestock care:

Behavioral Monitoring: Behavior analysis systems utilize a variety of sensor technologies, including accelerometers, gyroscopes, GPS trackers, and video cameras, to monitor the behavioral activities of livestock in real-time. Sensors are deployed in barns, pastures, and feeding areas to capture data on key behavioral indicators such as locomotion, feeding behavior, social interactions, rest patterns, and vocalizations. Real-time behavioral monitoring enables farmers to gain insights into the daily routines and social dynamics of livestock populations, facilitating early detection of deviations from normal behavior patterns that may

58

indicate health problems or environmental stressors.

AI-driven Analytics: AI algorithms process and analyze data streams from behavioral monitoring sensors to identify patterns, trends, and anomalies indicative of changes in animal behavior. Machine learning techniques, such as pattern recognition, clustering, and anomaly detection, enable AI algorithms to distinguish between normal and abnormal behavior patterns, classify behavioral states, and predict future behavioral responses based on historical data. AI-driven analytics provide farmers with actionable insights into animal behavior, enabling them to make informed decisions and implement targeted interventions to address emerging health or welfare concerns in livestock populations.

Health and Welfare Assessment: Behavior analysis systems enable farmers to assess the health and welfare status of livestock based on their behavioral responses to various stimuli and

environmental conditions. Changes in behavior, such as reduced activity levels, changes in posture, increased aggression, or altered feeding patterns, may indicate underlying health problems, pain, or discomfort in individual animals. By monitoring behavioral indicators in real-time, farmers can detect early signs of illness, injury, or stress, enabling timely interventions to mitigate risks and improve animal well-being.

Environmental Enrichment and Management: Behavior analysis informs the design and implementation of environmental enrichment strategies to promote optimal welfare and productivity in livestock populations. By understanding the behavioral needs and preferences of different species and breeds, farmers can design environments that support natural behaviors and encourage positive interactions among animals. Enriched environments may include features such as spacious living areas, access to pasture or outdoor space, comfortable resting areas,

and opportunities for socialization and exploration. Behavioral analysis helps farmers assess the effectiveness of environmental enrichment measures and make adjustments to optimize animal comfort and satisfaction.

Precision Livestock Management:

Behavior analysis facilitates precision livestock management practices by enabling farmers to tailor management interventions and resources to the specific needs and preferences of individual animals. By identifying behavioral traits and preferences, farmers can optimize feeding regimens, housing arrangements, and social groupings to maximize animal welfare and productivity. Precision livestock management strategies promote individualized care and attention, enhancing the health, comfort, and performance of livestock populations while minimizing resource wastage and environmental impact.

Soil Management

In "AI in Agriculture: How to Use AI in Farming," soil management emerges as a critical component of sustainable agricultural practices, with Artificial Intelligence (AI) technologies playing a pivotal role in optimizing soil health, fertility, and productivity. Soil management encompasses a range of practices aimed at preserving soil quality, preventing erosion, enhancing nutrient availability, and promoting sustainable crop growth. Through the integration of AI-driven analytics, remote sensing, and precision agriculture techniques, farmers can make informed decisions to improve soil management practices and maximize agricultural productivity. Here's how AI is transforming soil management:

Soil Health Assessment: AI technologies enable farmers to assess soil health and fertility through the analysis of soil samples, sensor data, and remote sensing imagery. Machine learning algorithms process soil data to identify key

indicators of soil health, such as organic matter content, nutrient levels, pH levels, and microbial activity. By analyzing historical soil data and environmental factors, AI models can detect trends, patterns, and anomalies that may impact soil productivity and resilience. Soil health assessment tools provide farmers with actionable insights into soil conditions, enabling them to implement targeted soil management strategies to improve fertility, structure, and nutrient cycling processes.

Precision Nutrient Management:

AI-driven nutrient management systems optimize fertilizer application rates and nutrient inputs based on real-time soil data, crop requirements, and environmental conditions. Machine learning algorithms analyze soil nutrient levels, crop uptake rates, and yield goals to develop customized nutrient management plans tailored to specific field conditions and crop needs. By optimizing nutrient application timing, placement, and formulation, farmers can minimize nutrient losses, reduce environmental pollution, and

63

maximize nutrient use efficiency, leading to improved crop yields and profitability while preserving soil health and water quality.

Soil Moisture Monitoring: AI-powered soil moisture monitoring systems track soil moisture levels across fields using sensors, satellite imagery, and weather data. Machine learning algorithms analyze moisture data to assess soil water-holding capacity, infiltration rates, and drainage characteristics, enabling farmers to optimize irrigation scheduling and water management practices. By monitoring soil moisture in real-time, farmers can prevent overwatering or underwatering, reduce water wastage, and mitigate risks of soil compaction, erosion, and nutrient leaching, promoting efficient water use and sustainable irrigation practices.

Soil Erosion Prediction and Control: AI technologies facilitate soil erosion prediction and control through the analysis of topographic data, rainfall

patterns, land use practices, and erosion modeling techniques. Machine learning algorithms predict erosion risk zones and identify vulnerable areas prone to soil loss and degradation. By analyzing historical erosion data and environmental factors, AI models can assess the effectiveness of erosion control measures, such as contour plowing, terracing, cover cropping, and vegetative buffer strips. Soil erosion prediction tools help farmers implement targeted erosion control strategies to protect soil resources, preserve crop productivity, and maintain environmental sustainability in agricultural landscapes.

Carbon Sequestration and Climate Mitigation: AI-driven soil management practices contribute to carbon sequestration and climate mitigation efforts by enhancing soil organic matter content and promoting carbon storage in agricultural soils. Machine learning algorithms analyze soil data and management practices to assess carbon sequestration potential and identify

opportunities for soil carbon enhancement. By adopting conservation tillage, crop rotation, cover cropping, and organic amendments, farmers can increase soil carbon stocks, reduce greenhouse gas emissions, and mitigate climate change impacts, while improving soil structure, fertility, and resilience to environmental stressors.

Soil Health Monitoring and Analysis

In "AI in Agriculture: How to Use AI in Farming," soil health monitoring and analysis stand as fundamental practices for sustainable agricultural management, where Artificial Intelligence (AI) technologies play a central role in revolutionizing traditional soil management approaches. Soil health monitoring involves the continuous assessment of key soil parameters to gauge fertility, structure, and microbial activity, while soil analysis encompasses the interpretation of soil data to inform management decisions and optimize agricultural practices. Here's how

AI transforms soil health monitoring and analysis:

Sensor Technologies: AI-enabled soil health monitoring systems leverage a variety of sensor technologies to capture data on critical soil parameters such as organic matter content, nutrient levels, pH levels, moisture content, and microbial activity. Sensors may include soil probes, spectrometers, moisture sensors, and microbial DNA analyzers deployed in fields or laboratory settings. Real-time data collection enables farmers to monitor soil conditions continuously and detect changes or trends over time, facilitating timely interventions to improve soil health and productivity.

AI-driven Analytics: AI algorithms process and analyze soil data to extract actionable insights and patterns that inform soil health assessment and management strategies. Machine learning techniques, such as regression analysis, clustering, and pattern recognition, enable AI models to identify correlations between

soil parameters, environmental factors, and agronomic practices. By analyzing historical soil data and environmental variables, AI algorithms can predict soil health trends, diagnose soil health problems, and recommend targeted interventions to improve soil fertility, structure, and microbial activity.

Remote Sensing and Imaging: AI-powered remote sensing technologies, including satellite imagery, aerial drones, and ground-based sensors, provide valuable data for soil health monitoring and analysis. Remote sensing platforms capture high-resolution images and spectral data of agricultural landscapes, enabling farmers to assess soil conditions, vegetation cover, and land use patterns from a bird's-eye view. AI algorithms process remote sensing data to map soil properties, identify soil degradation trends, and prioritize management interventions in areas of concern. Remote sensing technologies complement ground-based monitoring efforts and provide spatially

explicit information for precision soil management practices.

Predictive Modeling: AI-driven predictive modeling techniques forecast soil health outcomes and assess the effectiveness of soil management practices under different scenarios and management scenarios. Predictive models integrate soil data with climate data, crop performance data, and land management practices to simulate soil processes, predict soil fertility trends, and optimize soil management strategies. By simulating soil dynamics and predicting future soil health trajectories, predictive models enable farmers to evaluate the long-term impacts of management decisions, anticipate soil health risks, and adopt proactive measures to mitigate soil degradation and enhance soil resilience.

Decision Support Systems: AI-driven decision support systems provide farmers with tools and recommendations to guide soil health management decisions and practices. Decision support systems

integrate soil data, agronomic knowledge, and AI algorithms to generate personalized soil management plans tailored to specific field conditions and crop requirements. By considering soil health indicators, climate variability, and production goals, decision support systems help farmers optimize nutrient management, irrigation scheduling, crop rotations, and tillage practices to maximize soil health and productivity while minimizing environmental impact.

AI-driven Soil Nutrient Management

In "AI in Agriculture: How to Use AI in Farming," AI-driven soil nutrient management represents a transformative approach to optimizing soil fertility, improving crop yields, and promoting sustainable agricultural practices. Soil nutrient management involves the precise application of fertilizers and soil amendments to meet crop nutrient requirements while minimizing nutrient losses and environmental impact. Through

70

the integration of Artificial Intelligence (AI) technologies, farmers can enhance nutrient management practices by leveraging real-time data, predictive modeling, and decision support systems. Here's how AI-driven soil nutrient management is revolutionizing agriculture:

Soil Nutrient Analysis: AI technologies facilitate the analysis of soil nutrient levels through the interpretation of soil samples and sensor data. Machine learning algorithms process soil data to quantify nutrient concentrations, assess nutrient availability, and identify nutrient deficiencies or imbalances. By analyzing historical soil data and environmental factors, AI models can predict soil nutrient trends and recommend targeted nutrient management strategies to optimize soil fertility and crop nutrition.

Real-time Monitoring: AI-driven soil nutrient management systems enable farmers to monitor soil nutrient levels in real-time using sensors and remote sensing technologies. Sensors measure soil

parameters such as nitrogen, phosphorus, potassium, and micronutrient levels, providing immediate feedback on soil nutrient status. Remote sensing platforms capture spectral data of agricultural fields, enabling farmers to map spatial variations in soil fertility and prioritize nutrient management interventions. Real-time monitoring allows farmers to adjust nutrient application rates and timing based on current soil conditions, crop growth stage, and environmental factors, maximizing nutrient use efficiency and minimizing nutrient losses.

Predictive Modeling: AI-powered predictive modeling techniques forecast soil nutrient dynamics and assess the long-term impacts of nutrient management practices. Predictive models integrate soil data with climate data, crop performance data, and management practices to simulate soil nutrient cycling processes, predict nutrient availability, and optimize fertilizer application rates. By simulating different nutrient scenarios and management strategies, predictive models

72

help farmers evaluate the economic and environmental outcomes of nutrient management decisions and identify opportunities for improving soil health and productivity.

Decision Support Systems: AI-driven decision support systems provide farmers with tools and recommendations to guide nutrient management decisions and practices. Decision support systems integrate soil data, crop nutrient requirements, and AI algorithms to generate customized nutrient management plans tailored to specific field conditions and crop needs. By considering soil nutrient levels, crop uptake rates, and environmental factors, decision support systems help farmers optimize fertilizer application rates, formulations, and placement strategies to maximize crop yields while minimizing nutrient runoff, leaching, and pollution risks.

Variable Rate Application: AI technologies enable variable rate application (VRA) of fertilizers and soil

amendments based on spatial variations in soil nutrient levels and crop requirements. VRA systems use GPS technology and mapping algorithms to create prescription maps that guide automated fertilizer applicators to apply nutrients at variable rates across fields. By matching nutrient inputs to specific soil fertility zones and crop growth stages, VRA systems optimize nutrient use efficiency, reduce fertilizer wastage, and improve crop nutrient uptake, leading to more sustainable and cost-effective nutrient management practices.

Supply Chain Optimization

In "AI in Agriculture: How to Use AI in Farming," supply chain optimization emerges as a crucial aspect of modern agricultural management, where Artificial Intelligence (AI) technologies play a pivotal role in improving efficiency, reducing waste, and enhancing profitability throughout the agricultural value chain. Supply chain optimization in agriculture involves the coordination of various processes, including production, processing, distribution, and marketing, to ensure the timely delivery of high-quality products to consumers while minimizing costs and maximizing value. Here's how AI-driven supply chain optimization is transforming agriculture:

Demand Forecasting: AI-powered demand forecasting systems analyze historical sales data, market trends, weather patterns, and consumer preferences to predict future demand for agricultural products. Machine learning algorithms process large datasets to identify patterns, correlations, and seasonality in demand fluctuations, enabling farmers and agribusinesses to anticipate market trends and adjust production schedules accordingly. By accurately forecasting demand, AI technologies help prevent stockouts, minimize excess inventory, and optimize production planning, ensuring that agricultural products are available when and where they are needed.

Inventory Management: AI-driven inventory management systems optimize

the storage, tracking, and distribution of agricultural inputs and products throughout the supply chain. AI algorithms analyze inventory data, supplier lead times, production schedules, and demand forecasts to determine optimal inventory levels, reorder points, and safety stock thresholds. By automating inventory replenishment processes and implementing just-in-time inventory strategies, AI technologies help minimize carrying costs, reduce stockouts, and improve inventory turnover rates, leading to more efficient supply chain operations and improved cash flow management.

Logistics and Transportation: AI

technologies optimize logistics and transportation processes by optimizing route planning, vehicle scheduling, and load optimization. AI-driven routing algorithms analyze factors such as delivery locations, traffic patterns, fuel costs, and vehicle capacities to identify the most efficient routes and schedules for transporting agricultural products from farm to market. By minimizing

76

transportation costs and reducing delivery times, AI technologies help improve supply chain agility, enhance customer satisfaction, and reduce carbon emissions associated with transportation activities.

Quality Control and Traceability:

AI-powered quality control systems monitor product quality and safety throughout the supply chain, from farm to fork. AI algorithms analyze sensor data, imaging technologies, and quality metrics to detect defects, contamination, and spoilage in agricultural products. By implementing real-time quality control measures and automated inspection processes, AI technologies help ensure that only high-quality products reach consumers, reducing the risk of recalls, liability, and brand damage. Furthermore, AI-driven traceability systems enable stakeholders to track the origin, production methods, and handling practices of agricultural products, enhancing transparency, accountability, and food safety standards across the supply chain.

Market Intelligence and Pricing Optimization: AI technologies provide farmers and agribusinesses with market intelligence and pricing optimization tools to make informed decisions about pricing strategies, product positioning, and market segmentation. AI algorithms analyze market data, competitor pricing, consumer behavior, and economic indicators to identify pricing trends and opportunities. By dynamically adjusting prices and promotions based on demand elasticity and market conditions, AI-driven pricing optimization systems help maximize revenue, improve profit margins, and optimize resource allocation across the supply chain.

AI-driven Logistics and Distribution

In "AI in Agriculture: How to Use AI in Farming," AI-driven logistics and distribution represent transformative approaches to streamline the movement of agricultural products from farm to market, optimize supply chain operations, and enhance overall efficiency and profitability in the agricultural sector. Leveraging Artificial Intelligence (AI) technologies, farmers and agribusinesses can revolutionize logistics and distribution processes, ensuring timely delivery of high-quality products while minimizing costs and reducing environmental impact. Here's how AI-driven logistics and distribution are reshaping agriculture:

Route Optimization: AI-powered route optimization algorithms analyze factors such as delivery locations, traffic patterns, vehicle capacities, and fuel costs to determine the most efficient routes for transporting agricultural products. By considering real-time traffic data, weather conditions, and delivery constraints, AI algorithms dynamically adjust route plans to minimize transportation costs, reduce delivery times, and optimize resource utilization. Route optimization helps farmers and distributors maximize fleet efficiency, improve on-time delivery performance, and enhance customer

satisfaction while reducing fuel consumption and carbon emissions.

Vehicle Scheduling and Management: AI technologies optimize vehicle scheduling and management processes by dynamically allocating resources, coordinating deliveries, and optimizing vehicle utilization throughout the supply chain. AI-driven scheduling algorithms consider factors such as delivery priorities, vehicle capacities, driver availability, and service level agreements to optimize delivery schedules and allocate resources effectively. By automating scheduling tasks and optimizing load assignments, AI technologies help minimize idle time, reduce transportation costs, and improve overall fleet productivity, leading to more efficient logistics operations and enhanced customer service levels.

Predictive Maintenance: AI-driven predictive maintenance systems monitor vehicle health and performance using

sensor data, telematics, and predictive analytics techniques. Machine learning algorithms analyze vehicle diagnostic data, engine performance metrics, and historical maintenance records to identify patterns, trends, and anomalies indicative of potential equipment failures or maintenance issues. By predicting maintenance needs in advance, AI technologies enable fleet managers to schedule preventive maintenance activities, replace worn components, and address mechanical problems before they escalate, minimizing downtime, reducing repair costs, and prolonging vehicle lifespan.

Inventory Management and Warehousing: AI-powered inventory management systems optimize warehousing and storage operations by automating inventory tracking, optimizing stock levels, and streamlining order fulfillment processes. AI algorithms analyze inventory data, demand forecasts, and order patterns to determine optimal

inventory levels, reorder points, and storage locations. By implementing automated inventory replenishment strategies and real-time inventory tracking systems, AI technologies help reduce stockouts, minimize excess inventory, and improve inventory turnover rates, leading to more efficient warehousing operations and reduced carrying costs.

Last-Mile Delivery Optimization:

AI technologies optimize last-mile delivery operations by improving delivery routing, order consolidation, and customer communication processes. Last-mile delivery optimization algorithms consider factors such as delivery time windows, customer preferences, package sizes, and delivery vehicle capacities to optimize delivery routes and schedules. By prioritizing deliveries, optimizing delivery sequences, and providing real-time delivery updates to customers, AI-driven last-mile delivery solutions help reduce delivery times, enhance customer satisfaction, and improve overall delivery performance, leading to increased competitiveness and customer loyalty.

Predictive Analytics for Supply and Demand Matching

In "AI in Agriculture: How to Use AI in Farming," predictive analytics for supply and demand matching represents a critical capability empowered by Artificial Intelligence (AI) technologies. This approach revolutionizes how agricultural stakeholders manage production, distribution, and marketing efforts to align supply with fluctuating demand patterns in the marketplace. By leveraging AI-driven predictive analytics, farmers, distributors, and retailers can anticipate market trends, optimize inventory levels, and maximize profitability while minimizing waste and inventory obsolescence. Here's how predictive analytics for supply and demand matching transforms agriculture:

Market Trend Analysis: AI-powered predictive analytics systems analyze historical sales data, market trends, consumer behavior, and external factors such as weather patterns and economic indicators to identify emerging trends and patterns in supply and demand dynamics. By leveraging machine learning algorithms, these systems can detect subtle changes in consumer preferences, buying patterns, and seasonal demand fluctuations, enabling agricultural stakeholders to anticipate shifts in market demand and adjust production and inventory levels accordingly.

Demand Forecasting: Predictive analytics models use historical data and statistical algorithms to forecast future demand for agricultural products across different markets, regions, and product categories. By analyzing historical sales data, promotional activities, and market drivers, predictive models can predict demand trends with a high degree of accuracy, enabling farmers and agribusinesses to adjust production schedules, allocate resources, and optimize inventory levels to meet expected

demand while minimizing stockouts and excess inventory.

Inventory Optimization: AI-driven predictive analytics enable farmers, distributors, and retailers to optimize inventory levels and distribution networks based on predicted demand forecasts and market insights. By analyzing inventory data, lead times, and demand variability, predictive analytics models can identify optimal stocking levels, reorder points, and safety stock thresholds to ensure adequate supply while minimizing carrying costs and inventory obsolescence. By optimizing inventory management practices, agricultural stakeholders can improve cash flow, reduce storage costs, and enhance overall supply chain efficiency.

Production Planning and Scheduling: Predictive analytics systems assist farmers and agribusinesses in optimizing production planning and scheduling activities based on anticipated market demand and resource availability.

By analyzing historical yield data, crop performance metrics, and agronomic factors, predictive models can generate production forecasts, recommend planting schedules, and optimize resource allocation to maximize crop yields and meet market demand. By aligning production schedules with market requirements, agricultural stakeholders can minimize production costs, reduce waste, and improve overall profitability.

Pricing Strategy Optimization:
Predictive analytics models help agricultural stakeholders optimize pricing strategies and promotional activities based on predicted supply and demand dynamics and competitive market conditions. By analyzing pricing data, competitor pricing strategies, and consumer behavior, predictive models can identify pricing opportunities, recommend price adjustments, and optimize promotional campaigns to maximize revenue and profitability. By dynamically adjusting prices in response to changes in market conditions and consumer preferences,

agricultural stakeholders can improve price competitiveness, enhance customer loyalty, and drive sustainable revenue growth.

CHAPTER 3

CASE STUDIES AND EXAMPLES

In "AI in Agriculture: How to Use AI in Farming," case studies and examples serve as invaluable illustrations of how Artificial Intelligence (AI) technologies are transforming various aspects of agriculture, from crop management to supply chain logistics. These real-world examples demonstrate the practical applications and benefits of AI-driven solutions in improving productivity, sustainability, and profitability across different agricultural domains. Here are some case studies and examples highlighted in the book:

Precision Crop Management: The book features case studies of farmers and agronomists leveraging AI-driven precision agriculture technologies to optimize crop management practices. For example, a soybean farmer in the Midwest uses satellite imagery and machine learning algorithms to map field variability, identify areas of stress or nutrient deficiency, and adjust fertilizer applications accordingly. By adopting precision crop management techniques, the farmer achieves higher yields, reduced input costs, and improved environmental stewardship.

Predictive Analytics for Disease Management: The book showcases examples of how AI-powered predictive analytics are revolutionizing disease management strategies in agriculture. For instance, a vineyard owner in California uses sensor data and predictive modeling to forecast the risk of grapevine diseases such as powdery mildew and downy mildew. By monitoring environmental conditions, disease prevalence, and vineyard microclimates, the vineyard owner can implement targeted disease management interventions, such as fungicide applications or canopy management practices, to prevent disease outbreaks and protect crop health.

Supply Chain Optimization: Case studies highlight how AI-driven supply chain optimization solutions are improving efficiency and reducing costs throughout the agricultural value chain. For instance, a large-scale produce distributor uses AI-powered logistics platforms to optimize route planning, vehicle scheduling, and inventory management. By leveraging predictive analytics and real-time data, the distributor can minimize transportation costs, reduce delivery times, and optimize inventory levels, ensuring timely delivery of fresh produce to retail markets while minimizing waste and spoilage.

Livestock Monitoring and Management: The book features examples of how AI technologies are enhancing livestock monitoring and management practices on farms and ranches worldwide. For example, a dairy farmer implements AI-driven systems for automated health monitoring and behavioral analysis of dairy cows. By using wearable sensors and computer vision technology, the farmer can track individual cow activity levels, detect signs of illness or distress, and optimize feeding and milking routines to improve herd health and productivity.

Market Intelligence and Decision Support: The book highlights examples of how AI-powered market intelligence and decision support systems are empowering agricultural stakeholders to make informed decisions and capitalize on market opportunities. For instance, a grain trader uses AI algorithms to analyze commodity market data, weather forecasts, and geopolitical events to identify trading opportunities and hedge price risks. By leveraging predictive analytics and data-driven insights, the grain trader can optimize trading strategies, mitigate market volatility, and maximize profitability in commodity markets.

In summary, case studies and examples play a pivotal role in illustrating the diverse applications and benefits of AI in agriculture. By showcasing real-world success stories and practical examples of AI-driven solutions, the book provides readers with insights into how AI technologies are reshaping the future of farming and revolutionizing agricultural practices worldwide.

Real-world examples of AI implementation in agriculture

Real-world examples of AI implementation in agriculture abound, demonstrating the transformative potential of Artificial Intelligence (AI) technologies across various facets of farming. In "AI in Agriculture: How to Use AI in Farming," the book features compelling examples that showcase the practical applications and benefits of AI in modern agriculture. Here are some real-world examples highlighted in the book:

Weed Detection and Management: Farmers are employing AI-driven image recognition systems to identify and manage weeds in crop fields. For instance, a corn farmer utilizes drones

92

equipped with high-resolution cameras and AI algorithms to detect and classify weed species in real-time. By pinpointing weed-infested areas, the farmer can implement targeted herbicide applications or mechanical weed control measures, minimizing herbicide usage, reducing labor costs, and preserving crop yields.

Pest Monitoring and Control: AI

technologies are revolutionizing pest monitoring and control strategies in agriculture. For example, a fruit orchard manager utilizes smart traps equipped with sensors and AI algorithms to monitor pest populations and predict pest outbreaks. By analyzing trap data and environmental variables, such as temperature and humidity, the orchard manager can anticipate pest activity, implement timely interventions, and reduce reliance on chemical pesticides, promoting ecological balance and sustainable pest management practices.

Crop Disease Diagnosis: AI-driven

diagnostic tools are empowering farmers to

diagnose and manage crop diseases more effectively. For instance, a wheat farmer uses smartphone apps powered by AI algorithms to identify common wheat diseases based on leaf symptoms and environmental conditions. By capturing images of diseased leaves and submitting them to the app, the farmer receives instant diagnostic results and personalized recommendations for disease management strategies, such as fungicide applications or crop rotation practices, helping to mitigate disease spread and minimize yield losses.

Soil Health Assessment: Farmers are leveraging AI technologies to assess soil health and fertility, enabling data-driven decision-making in soil management practices. For example, a vegetable grower utilizes soil sensors and AI-driven analytics to monitor soil moisture, nutrient levels, and pH levels in real-time. By analyzing soil data and historical trends, the grower can optimize irrigation scheduling, adjust fertilizer applications, and improve soil structure and fertility, leading to healthier

94

crops, higher yields, and improved resource efficiency.

Predictive Yield Modeling:

AI-powered predictive modeling tools are transforming yield forecasting and production planning in agriculture. For instance, a grain farmer employs machine learning algorithms to analyze historical yield data, weather patterns, and agronomic practices to predict crop yields for the upcoming season. By simulating different scenarios and input variables, the farmer can optimize planting decisions, allocate resources efficiently, and mitigate production risks, enhancing profitability and resilience in the face of climatic uncertainties.

Success Stories and Challenges in Adopting AI Technology

In "AI in Agriculture: How to Use AI in Farming," success stories and challenges encountered in adopting AI technology in agriculture serve as valuable insights into the transformative potential and practical considerations associated with integrating AI solutions into farming practices. Here are some success stories and challenges highlighted in the book:

Success Stories:

Increased Productivity: Farmers and agribusinesses report significant improvements in productivity and efficiency following the adoption of AI-driven technologies. For example, a vineyard owner implements AI-powered predictive analytics to optimize irrigation scheduling and grapevine management practices, resulting in higher yields, improved fruit quality, and reduced water usage.

Cost Savings: Agricultural stakeholders realize cost savings and operational efficiencies through the adoption of AI technologies. For instance, a large-scale dairy farm implements AI-driven robotic milking systems to automate milking operations and monitor cow health, leading to labor savings, increased milk production, and improved animal welfare.

Enhanced Sustainability: AI-enabled precision agriculture practices promote sustainable farming methods and environmental stewardship. For example, a row crop farmer adopts AI-driven variable rate application (VRA) technology to optimize fertilizer and pesticide applications, reducing chemical usage, minimizing environmental impact, and improving soil health and water quality.

Challenges:

Access to Technology: One of the primary challenges in adopting AI technology in agriculture is access to affordable and accessible technology solutions, particularly for small-scale farmers and rural communities. Limited access to high-speed internet connectivity, technical support, and training programs may hinder the widespread adoption of AI-driven technologies in agricultural regions.

Data Privacy and Security: The collection and analysis of agricultural data raise concerns about data privacy and security. Farmers and agribusinesses must address data privacy regulations, data ownership issues, and cybersecurity risks associated with AI-enabled farming systems to protect sensitive information and maintain trust among stakeholders.

Technical Expertise and Training: The successful implementation of AI technologies requires specialized technical expertise and training. Farmers and agricultural professionals may face challenges in acquiring the necessary skills and knowledge to effectively deploy, manage, and troubleshoot AI-driven systems, highlighting the importance of education and training initiatives to support technology adoption in agriculture.

Integration and Interoperability: The integration of AI technologies with existing farm management systems and equipment presents challenges related to interoperability and compatibility. Farmers may encounter difficulties in integrating AI-driven solutions with legacy hardware and software systems, requiring investment in infrastructure upgrades and technical support services to ensure seamless integration and interoperability.

Return on Investment (ROI): The upfront costs associated with implementing AI technologies in agriculture may pose barriers to adoption, particularly for small and medium-sized farms with limited financial resources. Farmers and agribusinesses must carefully evaluate the potential return on investment (ROI) and long-term benefits of AI adoption, considering factors such as upfront capital costs, ongoing maintenance expenses, and expected improvements in productivity, profitability, and sustainability.

Lessons Learned from Various Applications across Different Regions

In "AI in Agriculture: How to Use AI in Farming," lessons learned from diverse applications across different regions offer valuable insights into the opportunities and challenges associated with implementing AI technologies in agriculture. Drawing from experiences around the world, these lessons shed light on best practices, key considerations, and innovative approaches to leveraging AI for sustainable and productive farming. Here are some lessons learned from various applications across different regions:

Contextual Adaptation: Agricultural applications of AI must be tailored to the specific needs, conditions, and challenges of different regions and farming systems. Lessons from diverse regions emphasize the importance of contextual adaptation and customization of AI solutions to account for factors such as climate variability, soil conditions, crop diversity, and socio-economic contexts. One-size-fits-all approaches may not be suitable for addressing the complex and heterogeneous nature of agricultural landscapes, highlighting the need for localized and contextually relevant AI solutions.

100

Collaboration and Knowledge Sharing: Successful AI applications in agriculture often involve collaboration and knowledge sharing among diverse stakeholders, including farmers, researchers, technology developers, and policymakers. Lessons from various regions underscore the importance of fostering partnerships, building networks, and sharing best practices to accelerate technology adoption, facilitate learning, and promote innovation in agriculture. Collaborative platforms, such as farmer cooperatives, research consortia, and digital extension services, play a crucial role in connecting stakeholders, facilitating knowledge exchange, and fostering a culture of continuous learning and improvement.

Capacity Building and Skills Development: The successful adoption and implementation of AI technologies in agriculture require investing in capacity building and skills development initiatives to empower farmers and agricultural

101

professionals with the knowledge, skills, and resources needed to harness the full potential of AI. Lessons from different regions highlight the importance of providing training programs, technical assistance, and hands-on learning opportunities to enable farmers to understand AI concepts, navigate digital tools, and integrate technology into their farming practices effectively. Building digital literacy and technical proficiency among agricultural stakeholders is essential for driving technology adoption and ensuring sustainable and inclusive development in rural communities.

Policy and Regulatory Frameworks:

The regulatory environment and policy frameworks significantly influence the adoption and deployment of AI technologies in agriculture. Lessons from various regions underscore the need for clear and supportive policy frameworks that promote innovation, facilitate technology transfer, and address regulatory challenges related to data privacy, intellectual property rights, and ethical considerations. Policymakers play a critical role in creating an enabling environment for technology adoption by establishing standards, incentives, and regulatory mechanisms that foster responsible and equitable use of AI in agriculture.

Continuous Monitoring and Evaluation: Monitoring and evaluation of AI applications in agriculture are essential for assessing impact, identifying opportunities for improvement, and refining strategies over time. Lessons from different regions emphasize the importance of collecting and analyzing data on technology adoption, performance metrics, and socio-economic outcomes to measure progress, inform decision-making, and drive evidence-based policy development. Continuous monitoring and evaluation enable stakeholders to identify success stories, address challenges, and adapt strategies in response to changing conditions, ensuring that AI investments

103

contribute to sustainable and inclusive agricultural development.

CHAPTER 4

IMPLEMENTING AI IN AGRICULTURE

In "AI in Agriculture: How to Use AI in Farming," the process of implementing AI technologies in agriculture involves several key steps and considerations to ensure successful integration and adoption. From defining objectives to evaluating outcomes, agricultural stakeholders must navigate various stages of implementation to harness the full potential of AI for sustainable and productive farming practices. Here's a guide to implementing AI in agriculture:

Define Objectives and Priorities: The first step in implementing AI in agriculture is to define clear objectives and priorities aligned with the needs and goals of farmers, agribusinesses, and rural communities. Stakeholders must identify specific challenges, opportunities, and desired outcomes that AI technologies can address, such as improving crop yields, reducing input costs, enhancing resource efficiency, or promoting environmental sustainability.

Assess Readiness and Capacity:

Agricultural stakeholders should assess their readiness and capacity to adopt and leverage AI technologies effectively. This involves evaluating factors such as technical infrastructure, digital literacy, financial resources, and organizational readiness to support technology adoption and implementation. Assessing readiness helps identify potential barriers, gaps, and opportunities for capacity building and technology integration.

Identify Suitable AI Solutions:

Once objectives are defined and readiness is assessed, stakeholders can explore and evaluate different AI solutions and technologies suitable for their specific needs and contexts. This may involve researching available AI tools, platforms,

105

and applications, consulting with technology providers, and seeking recommendations from peers and experts in the field. Identifying suitable AI solutions requires considering factors such as functionality, scalability, interoperability, and cost-effectiveness.

Pilot Testing and Validation:

Before full-scale deployment, it is advisable to conduct pilot testing and validation of selected AI solutions in real-world agricultural settings. Pilot projects allow stakeholders to assess the feasibility, performance, and usability of AI technologies in practical farming scenarios, identify potential challenges and limitations, and gather feedback from end-users. Pilot testing helps validate assumptions, refine strategies, and inform decision-making regarding technology adoption and scale-up.

Customize and Adapt Solutions:

AI solutions should be customized and adapted to the specific needs, conditions, and requirements of agricultural

106

stakeholders and farming systems. Customization may involve tailoring algorithms, interfaces, and user experiences to match local contexts, languages, and cultural preferences. Adapting solutions to local conditions and user feedback enhances usability, acceptance, and impact, increasing the likelihood of successful implementation and adoption.

Provide Training and Support:

Training and capacity building are essential components of successful AI implementation in agriculture. Agricultural stakeholders, including farmers, extension workers, and agribusiness professionals, should receive comprehensive training and technical support to build digital literacy, develop technical skills, and effectively utilize AI technologies in their daily operations. Training programs should be participatory, hands-on, and tailored to the needs and preferences of end-users, empowering them to maximize the benefits of AI for farming.

Monitor, Evaluate, and Iterate:

Continuous monitoring, evaluation, and iteration are integral to the successful implementation and refinement of AI technologies in agriculture. Stakeholders should establish monitoring and evaluation frameworks to track progress, measure impact, and identify areas for improvement throughout the implementation process. Regular feedback loops and performance assessments enable stakeholders to identify successes, address challenges, and adapt strategies in response to changing conditions and evolving needs.

Foster Collaboration and Knowledge Sharing:

Collaboration and knowledge sharing among diverse stakeholders are key drivers of successful AI implementation in agriculture. Stakeholders should foster partnerships, networks, and communities of practice to share experiences, lessons learned, and best practices related to AI adoption and integration. Collaborative platforms facilitate peer learning, innovation, and collective problem-solving, fostering a culture of continuous improvement and collective action in agriculture.

Steps to Integrate AI Solutions into Existing Farming Practices

In "AI in Agriculture: How to Use AI in Farming," integrating AI solutions into existing farming practices requires a systematic approach that considers the unique needs, contexts, and capabilities of agricultural stakeholders. By following these steps, farmers, agribusinesses, and rural communities can effectively harness the transformative potential of AI technologies to enhance productivity, sustainability, and resilience in farming:

Assess Current Practices and Challenges: Begin by conducting a comprehensive assessment of existing farming practices, challenges, and opportunities. Identify areas where AI technologies can address specific pain points or improve operational efficiency,

109

such as crop management, pest control, irrigation, or supply chain logistics. Engage stakeholders, including farmers, agronomists, and extension agents, in the assessment process to ensure a thorough understanding of local needs and priorities.

Define Objectives and Goals:

Clearly define objectives and goals for integrating AI solutions into farming practices. Determine key performance indicators (KPIs) and desired outcomes, such as increased crop yields, reduced input costs, improved resource efficiency, or enhanced environmental sustainability. Align objectives with broader farm management strategies and long-term sustainability goals to ensure coherence and alignment with overall farm objectives.

Identify Suitable AI Technologies:

Explore and evaluate a range of AI technologies and solutions tailored to the specific needs and contexts of farming operations. Consider factors such as functionality, scalability, interoperability, and cost-effectiveness when selecting AI solutions. Consult with technology providers, industry experts, and peer farmers to identify suitable AI tools, platforms, and applications that meet the requirements and preferences of your farm.

Conduct Pilot Projects and Proof-of-Concept Trials: Before full-scale implementation, conduct pilot projects and proof-of-concept trials to test the feasibility and performance of selected AI solutions in real-world farming conditions. Design pilot projects to address specific use cases or operational challenges, such as precision irrigation, crop monitoring, or pest detection. Gather feedback from end-users, monitor performance metrics, and assess the usability and effectiveness of AI technologies during the pilot phase.

Customize and Adapt Solutions: Customize and adapt AI solutions to match the unique needs, conditions, and preferences of your farming operation.

111

Tailor algorithms, interfaces, and user experiences to align with local contexts, languages, and cultural norms. Customize AI solutions to accommodate variations in soil types, climate conditions, crop varieties, and farming practices, ensuring relevance and effectiveness in diverse agricultural settings.

Provide Training and Capacity Building: Invest in training and capacity building initiatives to empower farmers and agricultural professionals with the knowledge, skills, and confidence to use AI technologies effectively. Develop training programs, workshops, and educational materials to build digital literacy, enhance technical proficiency, and foster adoption of AI solutions. Provide hands-on training and technical support to ensure that farmers can navigate AI tools, interpret data insights, and make informed decisions in their farming practices.

Establish Data Management and Governance Frameworks: Develop

robust data management and governance frameworks to ensure responsible and ethical use of AI technologies in farming. Establish protocols for data collection, storage, sharing, and privacy protection to safeguard sensitive information and uphold data integrity. Implement data governance policies that adhere to regulatory requirements, industry standards, and best practices for data security and confidentiality.

Monitor, Evaluate, and Iterate:

Continuously monitor and evaluate the performance and impact of AI solutions in farming practices. Track key performance indicators, such as yield improvements, cost savings, resource efficiencies, and environmental outcomes, to assess the effectiveness of AI interventions. Solicit feedback from end-users, stakeholders, and community members to identify areas for improvement and refinement. Iterate on AI solutions based on feedback and lessons learned to optimize performance and maximize value in farming operations.

Considerations for Selecting and Deploying AI Technologies

In "AI in Agriculture: How to Use AI in Farming," selecting and deploying AI technologies requires careful consideration of various factors to ensure successful implementation and integration into farming practices. By addressing these considerations, agricultural stakeholders can maximize the benefits of AI while mitigating risks and challenges. Here are key considerations for selecting and deploying AI technologies in agriculture:

Understand Farming Needs and Objectives:
Before selecting AI technologies, it is essential to understand the specific needs, challenges, and objectives of farming operations. Identify areas where AI can add value, such as crop management, pest control, irrigation optimization, or supply chain logistics. Align AI initiatives with broader farm

management strategies and sustainability goals to ensure coherence and relevance.

Evaluate Technical Requirements and Capabilities:

Assess the technical requirements and capabilities needed to deploy and operate AI technologies effectively. Consider factors such as hardware compatibility, data connectivity, computational resources, and technical expertise required to support AI applications. Ensure that farming infrastructure and IT systems can support the deployment and integration of AI solutions.

Consider Data Accessibility and Quality:

Data accessibility and quality are critical considerations for AI deployment in agriculture. Evaluate the availability, reliability, and granularity of data sources, such as satellite imagery, weather data, soil maps, and sensor measurements. Assess data quality, completeness, and consistency to ensure that AI algorithms can generate accurate and reliable insights. Address data gaps and limitations through data collection, validation, and integration efforts.

Assess Scalability and Adaptability: Select AI technologies that are scalable and adaptable to evolving farming needs and conditions. Consider scalability factors such as data volume, processing speed, and model complexity to accommodate growing data requirements and expanding operations. Choose flexible and modular AI solutions that can be customized and adapted to different crops, regions, and farming practices.

Evaluate Cost-Benefit Considerations: Conduct a thorough cost-benefit analysis to evaluate the financial implications of AI deployment in agriculture. Consider upfront costs such as technology acquisition, hardware infrastructure, software licensing, and implementation services. Assess potential benefits, such as yield improvements, cost savings, resource efficiencies, and

environmental outcomes, to determine the return on investment (ROI) and long-term viability of AI initiatives.

Address Ethical and Regulatory Considerations: Consider ethical and regulatory considerations associated with AI deployment in agriculture. Ensure compliance with data privacy regulations, intellectual property rights, and ethical guidelines governing AI development and deployment. Address concerns related to data ownership, consent, transparency, and accountability to build trust and confidence among stakeholders.

Foster Stakeholder Engagement and Collaboration: Involve key stakeholders, including farmers, agronomists, researchers, technology providers, and policymakers, in the selection and deployment of AI technologies. Foster collaboration and knowledge sharing among diverse stakeholders to ensure buy-in, shared understanding, and collective ownership of

AI initiatives. Encourage open communication, feedback, and participation to promote inclusive decision-making and stakeholder engagement.

Plan for Training and Capacity Building:
Invest in training and capacity building initiatives to empower farmers and agricultural professionals with the knowledge, skills, and confidence to use AI technologies effectively. Provide hands-on training, technical support, and educational resources to build digital literacy, enhance technical proficiency, and foster adoption of AI solutions. Tailor training programs to address specific needs, preferences, and learning styles of end-users.

Establish Monitoring and Evaluation Frameworks:
Develop robust monitoring and evaluation frameworks to track the performance and impact of AI technologies in farming practices. Define key performance indicators (KPIs) and metrics to measure progress, assess outcomes, and identify

areas for improvement. Implement regular performance assessments, data audits, and impact evaluations to ensure that AI initiatives deliver expected results and generate value for farming operations.

Addressing Challenges such as Cost, Infrastructure, and Data Privacy

In "AI in Agriculture: How to Use AI in Farming," overcoming challenges such as cost, infrastructure limitations, and data privacy concerns is crucial for the successful adoption and implementation of AI technologies in agriculture. By addressing these challenges proactively, agricultural stakeholders can unlock the transformative potential of AI while mitigating risks and ensuring sustainable and responsible technology deployment. Here's how to address these challenges in the book:

Cost Considerations:

Cost-Benefit Analysis: Conduct a thorough cost-benefit analysis to assess

119

the financial implications of adopting AI technologies in agriculture. Evaluate upfront costs associated with technology acquisition, hardware infrastructure, software licensing, and implementation services against potential benefits such as yield improvements, cost savings, and resource efficiencies.

Identify ROI Opportunities: Identify opportunities to generate return on investment (ROI) through AI-driven efficiencies, productivity gains, and revenue enhancement. Prioritize AI initiatives with the highest potential for delivering measurable and impactful outcomes while managing costs effectively.

Explore Funding Opportunities: Explore funding opportunities, grants, subsidies, and incentive programs available to support AI adoption and innovation in agriculture. Seek partnerships with government agencies, research institutions, and industry stakeholders to access funding sources and financial resources for technology investment.

Infrastructure Limitations:

Assess Technical Requirements: Assess the technical requirements and capabilities needed to support the deployment and operation of AI technologies in agriculture. Evaluate existing infrastructure, such as data connectivity, computational resources, and IT systems, to identify potential limitations and constraints.

Invest in Infrastructure Upgrades: Invest in infrastructure upgrades and technology enhancements to address gaps and limitations in existing farming infrastructure. Upgrade hardware components, expand network connectivity, and optimize IT systems to support the demands of AI-driven applications and data processing.

Leverage Cloud Computing: Leverage cloud computing and distributed computing platforms to access scalable and cost-

121

effective computing resources for AI deployment. Cloud-based solutions offer flexibility, scalability, and on-demand access to computing power, enabling agricultural stakeholders to overcome infrastructure limitations and accelerate technology adoption.

Data Privacy Concerns:

Establish Data Governance Frameworks: Establish robust data governance frameworks to ensure responsible and ethical use of agricultural data in AI applications. Develop policies, procedures, and protocols for data collection, storage, sharing, and privacy protection to safeguard sensitive information and uphold data integrity.

Implement Data Encryption and Security Measures: Implement data encryption, access controls, and security measures to protect agricultural data from unauthorized access, disclosure, or misuse. Encrypt sensitive data during transmission and storage, enforce access

122

controls, and monitor data activities to detect and mitigate security threats.

Enhance Transparency and Consent:

Enhance transparency and accountability in data practices by providing clear information and obtaining informed consent from stakeholders regarding data collection, use, and sharing activities. Communicate openly about data privacy policies, data rights, and data management practices to build trust and confidence among stakeholders.

CHAPTER 5

FUTURE TRENDS AND DEVELOPMENTS IN AGRICULTURE

In "AI in Agriculture: How to Use AI in Farming," exploring future trends and developments is essential to understanding the evolving landscape of agricultural technology and the potential impact of AI on farming practices. As technology continues to advance and new innovations emerge, several key trends are shaping the future of agriculture:

Advancements in Precision Agriculture: Precision agriculture is poised to continue evolving with the integration of AI-driven technologies. Future developments may include the widespread adoption of autonomous vehicles, drones, and robotics for precision planting, spraying, and harvesting. AI-powered sensors and imaging technologies will enable real-time monitoring of crop health, soil conditions, and environmental variables, facilitating data-driven decision-making and resource optimization.

Expansion of Digital Farming Solutions: Digital farming solutions will play an increasingly prominent role in modern agriculture, driven by advancements in AI, IoT (Internet of Things), and big data analytics. Integrated digital platforms and farm management software will enable farmers to collect, analyze, and visualize data from diverse sources, such as sensors, satellites, and weather stations, to optimize farm operations, manage risk, and enhance productivity.

Adoption of AI-driven Plant Breeding and Genetics: AI technologies are revolutionizing plant breeding and genetics by accelerating the development of new crop varieties with improved traits, such as yield potential, disease resistance, and stress tolerance.

125

AI algorithms can analyze genomic data, phenotype data, and environmental data to predict genotype-phenotype relationships and identify desirable traits for crop improvement. This could lead to the development of climate-resilient and high-yielding crop varieties tailored to specific agroecological regions and production systems.

Integration of AI into Supply Chain Management:

AI will increasingly be integrated into supply chain management processes to optimize logistics, distribution, and market intelligence in agriculture. AI-powered predictive analytics and demand forecasting tools will enable more efficient inventory management, transportation planning, and market optimization, reducing waste and improving market access for farmers and agribusinesses.

Emphasis on Sustainable and Regenerative Agriculture:

There is growing recognition of the importance of

sustainable and regenerative agriculture practices in addressing environmental challenges and promoting food security. AI technologies will play a critical role in supporting sustainable agriculture initiatives by optimizing resource use, minimizing environmental impact, and enhancing ecosystem services. AI-driven solutions for soil health management, water conservation, and biodiversity conservation will enable farmers to adopt more sustainable and resilient farming practices.

Integration of Blockchain and Traceability Solutions: Blockchain technology and traceability solutions will enhance transparency, traceability, and trust in agricultural supply chains. By leveraging blockchain-based platforms and distributed ledger technology, stakeholders can track the provenance and authenticity of agricultural products from farm to fork, ensuring food safety, quality, and sustainability. AI algorithms can analyze blockchain data to identify patterns, detect

127

anomalies, and optimize supply chain processes.

Expansion of AgriTech Startups and Innovation Ecosystems: The AgriTech startup ecosystem is expected to continue growing, fueled by investment, entrepreneurship, and technological innovation. Startups and innovators are developing disruptive AI-driven solutions for various agricultural applications, including crop monitoring, pest management, and farm robotics. Collaboration and partnerships between startups, research institutions, and industry stakeholders will drive innovation and foster a culture of entrepreneurship and creativity in agriculture.

Emerging AI Technologies Poised to Transform Agriculture

In "AI in Agriculture: How to Use AI in Farming," the integration of emerging AI technologies holds immense potential to revolutionize farming practices, enhance productivity, and address global food security challenges. As agricultural stakeholders continue to innovate and explore new frontiers in technology, several emerging AI technologies are poised to transform agriculture:

Quantum Computing: Quantum computing has the potential to revolutionize agricultural data analysis and computational modeling by exponentially increasing processing power and enabling complex simulations and optimizations. Quantum algorithms can solve optimization problems, such as crop planning, supply chain logistics, and genetic optimization, with unprecedented speed and efficiency, unlocking new insights and possibilities for agricultural innovation.

Federated Learning: Federated learning is a decentralized machine learning approach that enables model training across distributed devices and data sources while preserving data privacy and security. In agriculture, federated learning can empower farmers to

129

collaboratively train AI models using decentralized data from diverse farm locations and environments. This enables the development of localized and contextually relevant AI solutions for precision agriculture, pest management, and soil health monitoring.

Edge Computing: Edge computing brings AI processing capabilities closer to the source of data generation, enabling real-time analysis and decision-making at the edge of the network. In agriculture, edge computing devices, such as sensors, drones, and smart machinery, can leverage AI algorithms to process data locally and provide actionable insights to farmers in remote or resource-constrained environments. Edge AI solutions enhance efficiency, reduce latency, and enable autonomous decision-making in farm operations.

Generative Adversarial Networks (GANs): Generative adversarial networks (GANs) are AI models capable of generating synthetic data and simulating realistic scenarios by learning from real-world examples. In agriculture, GANs can be used to generate synthetic images of crops, pests, and diseases, facilitating data augmentation, model training, and scenario analysis. GANs enable farmers to simulate crop growth, predict pest outbreaks, and optimize resource allocation in virtual environments, enhancing decision-making and risk management.

Explainable AI (XAI): Explainable AI (XAI) is a critical area of research focused on developing AI models that are transparent, interpretable, and explainable to end-users. In agriculture, XAI techniques enable farmers to understand and trust AI-driven recommendations, predictions, and decisions by providing insights into model behavior and decision-making processes. XAI enhances transparency, accountability, and user acceptance of AI technologies in farming practices, fostering trust and confidence among stakeholders.

Synthetic Biology and Bioinformatics: Synthetic biology and bioinformatics leverage AI and computational techniques to design and engineer biological systems for agricultural applications. AI-driven bioinformatics tools enable the analysis of genomic data, gene expression patterns, and metabolic pathways to improve crop traits, enhance disease resistance, and optimize agronomic performance. Synthetic biology techniques, such as genome editing and gene sequencing, enable precision breeding and genetic engineering of crops for desired traits and characteristics.

Neuromorphic Computing:
Neuromorphic computing emulates the structure and functionality of the human brain to perform cognitive tasks and pattern recognition tasks efficiently. In agriculture, neuromorphic computing architectures can mimic biological neural networks and process sensory data from farm sensors, drones, and satellites in real-time. Neuromorphic AI systems enable

intelligent decision-making, adaptive learning, and autonomous control in agricultural robots, drones, and smart devices, enhancing efficiency and productivity in farming operations.

Potential Impact of AI on the Future of Farming and Food Production

In "AI in Agriculture: How to Use AI in Farming," the integration of AI technologies holds immense promise for revolutionizing farming practices and transforming the future of food production. As agricultural stakeholders continue to innovate and adopt AI-driven solutions, several potential impacts are poised to reshape the agricultural landscape:

Increased Productivity and Efficiency: AI technologies enable farmers to optimize resource use, improve decision-making, and enhance operational efficiency across the agricultural value chain. From precision agriculture and crop management to supply chain logistics and

133

market forecasting, AI-driven solutions streamline processes, reduce waste, and increase productivity, leading to higher yields and improved profitability for farmers.

Enhanced Sustainability and Environmental Stewardship: AI facilitates sustainable farming practices by enabling data-driven insights into soil health, water management, and crop nutrition. AI-powered sensors, drones, and satellite imagery enable real-time monitoring of environmental variables, empowering farmers to adopt precision irrigation, targeted pest control, and regenerative agriculture techniques. By optimizing resource use and minimizing environmental impact, AI contributes to sustainable and resilient food systems that promote biodiversity conservation and climate resilience.

Improved Crop Quality and Resilience: AI technologies enable breeders and geneticists to accelerate

crop improvement efforts and develop resilient and high-yielding crop varieties. AI-driven genomic analysis, phenotypic prediction, and trait optimization enable precision breeding and genetic engineering of crops with desirable traits, such as disease resistance, drought tolerance, and nutritional value. Enhanced crop quality and resilience contribute to food security, nutrition, and agricultural livelihoods, particularly in regions vulnerable to climate change and resource constraints.

Enhanced Market Access and Supply Chain Optimization: AI facilitates market access and supply chain optimization by enabling predictive analytics, demand forecasting, and inventory management in agriculture. AI-powered platforms and algorithms optimize distribution networks, logistics operations, and market intelligence, enabling farmers to access new markets, reduce post-harvest losses, and improve price transparency and negotiation power. By connecting producers with consumers and

streamlining value chain operations, AI promotes market efficiency, competitiveness, and inclusivity in agricultural trade.

Empowerment of Smallholder Farmers and Rural Communities:

AI technologies empower smallholder farmers and rural communities by providing access to information, resources, and decision support tools that enhance agricultural productivity and resilience. AI-driven extension services, mobile applications, and digital platforms enable farmers to access weather forecasts, market prices, agronomic advice, and financial services, fostering entrepreneurship, innovation, and economic empowerment in rural areas. By bridging the digital divide and promoting inclusive access to technology, AI contributes to poverty reduction, food security, and sustainable development in farming communities.

Accelerated Innovation and Entrepreneurship: AI fosters innovation and entrepreneurship in agriculture by enabling the development of new technologies, business models, and value-added services that address emerging challenges and opportunities in the food system. AI-driven startups, research institutions, and technology hubs drive innovation in areas such as agriTech, agroinformatics, and agripreneurship, catalyzing economic growth, job creation, and technology diffusion in rural and urban areas. By fostering a culture of innovation and collaboration, AI fuels creativity, resilience, and adaptability in the agriculture sector, shaping the future of farming and food production in a rapidly changing world.

Ethical and Societal Considerations of AI Adoption in Agriculture

In "AI in Agriculture: How to Use AI in Farming," the adoption of AI technologies in agriculture raises important ethical and societal considerations that must be carefully addressed to ensure responsible and equitable deployment. As agricultural stakeholders navigate the complexities of AI adoption, it is essential to prioritize ethical principles, promote social responsibility, and safeguard the well-being of individuals and communities. Here are key ethical and societal considerations of AI adoption in agriculture:

Data Privacy and Security:

Agricultural data, including farm management records, environmental data, and personal information, must be handled with utmost care to protect privacy and ensure data security. AI-driven agricultural systems collect and analyze vast amounts of sensitive data, raising concerns about data ownership, consent, and protection against unauthorized access or misuse. Stakeholders must implement robust data governance frameworks, encryption protocols, and access controls to safeguard data privacy and security throughout the data lifecycle.

138

Transparency and Accountability:

Transparency and accountability are fundamental principles guiding responsible AI adoption in agriculture. Farmers and end-users should have access to clear and understandable information about AI-driven technologies, including their capabilities, limitations, and potential impacts on farming practices and livelihoods. AI algorithms and decision-making processes should be transparent and explainable, enabling farmers to understand how AI recommendations are generated and make informed decisions about technology adoption.

Bias and Fairness:

AI algorithms and models may exhibit biases or discriminatory patterns that perpetuate inequalities and exacerbate social disparities in agriculture. Bias can arise from historical data, algorithmic design choices, or societal biases embedded in training data sets. Agricultural stakeholders must mitigate bias and promote fairness in AI systems by ensuring diverse representation in data collection, model training, and decision-making processes. Regular audits, bias assessments, and fairness evaluations can help identify and address bias in AI-driven solutions.

Human-Centric Design: Human-centric design principles prioritize the needs, values, and experiences of end-users in the development and deployment of AI technologies. Agricultural solutions should be designed with the user in mind, considering factors such as usability, accessibility, and cultural relevance. Engaging farmers, agronomists, and rural communities in the design process fosters user empowerment, trust, and acceptance of AI technologies, leading to more meaningful and impactful outcomes in farming practices.

Economic Implications and Employment: The adoption of AI technologies in agriculture may have significant economic implications for rural

livelihoods, employment opportunities, and income distribution. While AI-driven automation can improve productivity and reduce labor costs, it may also disrupt traditional farming practices, displace workers, and exacerbate rural-urban disparities. Agricultural stakeholders must anticipate and address the socio-economic impacts of AI adoption, including job displacement, skills development, and income redistribution, to ensure equitable access to the benefits of technology innovation.

Environmental and Ethical Sustainability:

AI adoption in agriculture should prioritize environmental sustainability, biodiversity conservation, and ethical stewardship of natural resources. AI-driven farming practices must minimize environmental impact, reduce greenhouse gas emissions, and promote soil health, water conservation, and biodiversity conservation. Sustainable agriculture principles, such as regenerative farming, agroecology, and organic farming, should guide AI deployment to mitigate environmental degradation and promote resilience in farming systems.

Governance and Regulation:

Effective governance and regulation are essential for ensuring responsible AI adoption and mitigating potential risks and harms in agriculture. Governments, regulatory agencies, and international organizations play a crucial role in developing and enforcing policies, standards, and guidelines for AI deployment in agriculture. Regulatory frameworks should address ethical, legal, and social implications of AI adoption, including data privacy, algorithmic accountability, liability, and intellectual property rights, to promote trust, accountability, and responsible innovation in agriculture.

CHAPTER 6

CONCLUSION

In "AI in Agriculture: How to Use AI in Farming," we have explored the transformative potential of artificial intelligence (AI) in revolutionizing farming practices and shaping the future of agriculture. From precision agriculture and crop management to supply chain optimization and market intelligence, AI technologies offer unprecedented opportunities to enhance productivity, sustainability, and resilience in farming operations.

Throughout this book, we have delved into the diverse applications of AI in agriculture, ranging from predictive analytics and precision farming to livestock monitoring and soil management. We have examined the fundamental principles of AI technology, explored emerging trends and developments, and discussed practical considerations for selecting, deploying, and integrating AI solutions into existing farming practices.

As we reflect on the implications of AI adoption in agriculture, it is essential to recognize the ethical, societal, and environmental considerations that accompany technological innovation. The responsible deployment of AI requires a commitment to data privacy, transparency, fairness, and human-centric design, ensuring that AI-driven solutions benefit farmers, consumers, and communities while minimizing risks and disparities.

Looking ahead, the future of agriculture is shaped by innovation, collaboration, and sustainability. By embracing AI technologies, agricultural stakeholders can address pressing challenges such as climate change, food security, and rural development, while unlocking new opportunities for innovation, entrepreneurship, and inclusive growth.

As we embark on this journey towards a more sustainable and resilient food system, let us remain vigilant in upholding ethical principles, promoting social responsibility, and safeguarding the well-being of people and planet. Together, we can harness the transformative power of AI to build a brighter and more prosperous future for agriculture, ensuring that farming remains not only a livelihood but also a beacon of hope for generations to come.

In closing, "AI in Agriculture: How to Use AI in Farming" serves as a roadmap for navigating the complexities of AI adoption in agriculture, offering practical guidance, insights, and best practices for harnessing the full potential of AI to drive positive change in farming and food production.

Thank you for embarking on this journey with us, and may your endeavors in agriculture be guided by innovation, sustainability, and shared prosperity.

Recap of Key Insights and Findings from "AI in Agriculture: How to Use AI in Farming"

Throughout "AI in Agriculture: How to Use AI in Farming," we've explored the transformative potential of artificial intelligence (AI) in revolutionizing farming practices and shaping the future of agriculture. Here's a recap of the key insights and findings from the book:

Applications of AI in Agriculture:

We've examined a wide range of applications of AI in agriculture, including precision agriculture, crop management, livestock monitoring, soil health analysis, supply chain optimization, and predictive analytics. AI-driven solutions enable farmers to optimize resource use, enhance decision-making, and improve productivity across the agricultural value chain.

Fundamentals of AI Technology:

We've explored the fundamental principles

of AI technology, including machine learning, computer vision, robotics, and natural language processing. Understanding the capabilities and limitations of AI algorithms is crucial for selecting, deploying, and integrating AI solutions into farming practices effectively.

Integration of AI into Agricultural Systems: We've discussed how AI integrates with existing agricultural systems, leveraging data from sensors, drones, satellites, and other sources to generate actionable insights and recommendations for farmers. By harnessing the power of AI, agricultural stakeholders can optimize inputs, reduce waste, and maximize yields while minimizing environmental impact.

Challenges and Considerations: We've identified key challenges and considerations associated with AI adoption in agriculture, including data privacy, infrastructure limitations, bias and fairness, economic implications, and

147

regulatory compliance. Addressing these challenges requires a proactive approach to ensure responsible and equitable deployment of AI technologies.

Ethical and Societal Considerations: We've highlighted the ethical and societal considerations of AI adoption in agriculture, emphasizing the importance of data privacy, transparency, fairness, human-centric design, and environmental sustainability. Promoting responsible innovation and social responsibility is essential for building trust and confidence in AI-driven solutions.

Future Trends and Developments: We've explored emerging trends and developments in AI technology, including quantum computing, federated learning, edge computing, synthetic biology, and neuromorphic computing. These technologies have the potential to revolutionize farming practices, enhance sustainability, and promote resilience in agricultural systems.

Conclusion and Outlook: In conclusion, "AI in Agriculture: How to Use AI in Farming" serves as a guidebook for navigating the complexities of AI adoption in agriculture, offering practical insights, strategies, and best practices for harnessing the full potential of AI to drive positive change in farming and food production. Looking ahead, the future of agriculture is shaped by innovation, collaboration, and sustainability, and by embracing AI technologies, agricultural stakeholders can address global challenges and unlock new opportunities for innovation and growth.

Final Thoughts on the Role of AI in Shaping the Future of Agriculture

As we conclude "AI in Agriculture: How to Use AI in Farming," it's evident that artificial intelligence (AI) holds tremendous promise for shaping the future of agriculture in profound and transformative ways. The integration of AI technologies into farming practices represents a pivotal moment in the evolution of agriculture, offering unprecedented opportunities to address global challenges, drive innovation, and build resilient and sustainable food systems.

AI has the potential to revolutionize every aspect of agriculture, from precision farming and crop management to supply chain optimization and market intelligence. By harnessing the power of AI, farmers can optimize resource use, enhance decision-making, and improve productivity while minimizing environmental impact and promoting biodiversity conservation.

However, the journey towards AI-enabled agriculture is not without its challenges and complexities. Ethical, societal, and environmental considerations must be carefully navigated to ensure responsible and equitable deployment of AI technologies. Data privacy, transparency, fairness, and human-centric design principles are essential pillars of responsible innovation, fostering trust, accountability, and inclusivity in AI-driven farming practices.

Looking ahead, the future of agriculture is shaped by innovation, collaboration, and sustainability. By embracing AI technologies, agricultural stakeholders can unlock new opportunities for innovation, entrepreneurship, and inclusive growth, driving positive change in farming and food production.

Call to Action: Embracing AI Technologies for Sustainable Development in Agriculture

As we conclude "AI in Agriculture: How to Use AI in Farming," it is imperative that stakeholders in the agriculture sector heed the call to embrace AI technologies for sustainable development. The integration of AI holds the potential to revolutionize farming practices, drive innovation, and foster resilience in agricultural systems. Here is a call to action for stakeholders:

Embrace Innovation: Let us embrace a culture of innovation and

151

entrepreneurship in agriculture, recognizing the transformative potential of AI technologies to address pressing challenges and unlock new opportunities for sustainable development. By fostering a spirit of curiosity, creativity, and collaboration, we can drive meaningful change and build a brighter future for agriculture.

Invest in Education and Training:

Let us invest in education and training programs that empower farmers, agronomists, and rural communities with the knowledge, skills, and resources needed to harness the power of AI technologies effectively. By providing access to training opportunities, technical support, and capacity-building initiatives, we can empower stakeholders to adopt AI-driven solutions and drive positive change in farming practices.

Foster Collaboration and Partnerships:

Let us foster collaboration and partnerships between

governments, industry stakeholders, research institutions, and civil society organizations to promote responsible innovation and technology transfer in agriculture. By leveraging collective expertise, resources, and networks, we can co-create innovative solutions, share best practices, and address shared challenges collaboratively.

Prioritize Sustainability and Resilience:
Let us prioritize sustainability, resilience, and environmental stewardship in AI-driven agriculture, ensuring that technology deployment aligns with principles of resource efficiency, biodiversity conservation, and climate resilience. By adopting regenerative farming practices, reducing greenhouse gas emissions, and promoting ecosystem health, we can build more resilient and adaptive agricultural systems that support sustainable development goals.

Advocate for Ethical and Inclusive AI: Let us advocate for ethical and inclusive AI policies, guidelines, and standards that promote transparency, fairness, accountability, and human-centric design in agricultural applications. By prioritizing data privacy, equity, and social responsibility, we can build trust and confidence in AI technologies and ensure that benefits are shared equitably among all stakeholders.

Drive Policy and Regulatory Reform: Let us drive policy and regulatory reform to create an enabling environment for AI adoption in agriculture, removing barriers to technology access, fostering innovation, and promoting responsible deployment. By advocating for evidence-based policies, regulatory frameworks, and incentives that support technology innovation and entrepreneurship, we can accelerate the transition towards sustainable and inclusive agricultural development.

154

Amplify Voices of Change: Let us amplify the voices of farmers, rural communities, and marginalized groups who are often most affected by agricultural challenges and technological changes. By engaging diverse stakeholders in decision-making processes, listening to their needs and concerns, and co-designing inclusive solutions, we can ensure that AI technologies serve the interests of all and leave no one behind.